Frequently ~~asked~~

all about
herbs

HYLA CASS, MD

Hyla Cass, M.D.
1608 Michael La, Pac Palisades, Ca 90272
(310) 459-9866 Fax (310) 459-9466
Hyla@cassmd.com • www.cassmd.com

The information contained in this book is based upon the research and personal and professional experiences of the author. It is not intended as a substitute for consulting with your physician or other health care provider. Any attempt to diagnose and treat an illness should be done under the direction of a health care professional.

The publisher does not advocate the use of any particular health care protocol, but believes the information in this book should be available to the public. The publisher and author are not responsible for any adverse effects or consequences resulting from the use of any of the suggestions, preparations, or procedures discussed in this book. Should the reader have any questions concerning the appropriateness of any procedure or preparation mentioned, the author and the publisher strongly suggest consulting a professional health care advisor.

Series cover designer: Eric Macaluso
Cover image courtesy of Steven Foster Group, Inc.

Avery Publishing Group, Inc.
120 Old Broadway, Garden City Park, NY 11040
1-800-548-5757 or visit us at www.averypublishing.com

ISBN: 0-89529-938-0

Printed in the United States of America

10   9   8   7   6   5   4

Contents

Introduction

People in every culture and every country have been using herbs as natural remedies for thousands of years. Safe and effective, herbs have passed the test of time and have helped millions of people maintain or recover their health.

Today, herbs have become the latest "discovery" in the United States and are being used by increasing numbers of people seeking natural alternatives to synthetic and often hazardous drugs. It may seem strange that modern drugs can be replaced with Grandma's teas, powders, and tinctures. But extensive scientific evidence supports the use of herbal medicines—and many modern drugs are actually synthetic versions of substances found in plants.

For years, American doctors have been dismissing the value of herbs, preferring instead to prescribe expensive drugs. European doctors, on the other hand, have always maintained their interest and skill in prescribing herbal medicines, or phytomedicines, in conjunction with pharmaceuticals.

Europeans have a stronger history of herb use compared with the United States, and, not surprisingly, most of the scientific research on herbs has been conducted in Europe.

Herbs possess many properties that make them superior to synthetic drugs, including being more in tune with the body and, therefore, creating fewer, and certainly less harmful, side effects. Most herbs also have multiple uses and side benefits. For example, ginkgo can improve memory, but it can also boost blood circulation to the arms and legs. In addition, many herbs easily surpass drugs in effectiveness. To wit, milk thistle, when used early enough, can protect the liver from mushroom poisoning.

Herbs are also appealing in that they can be purchased without a prescription, are relatively inexpensive, and are quite safe when used as directed. And that gets to the heart of *All About Herbs*. In this basic guide to herbal medicine, you'll learn what herbs are, how they work, and how to buy and use them. The other chapters will describe eight of the most popular herbal medicines and explain how you can use them to maintain the health of your body and mind. Whether you're young or elderly, a man or a woman, you will learn how to use herbs to help yourself feel better.

1.

Herbal Medicine From Folklore to Science

The use of herbs is as old as human history. From the most primitive jungle-dwellers to the highly sophisticated Chinese and Indians, every culture has had its herbal remedies.

Q. What is an herb?

A. A strict definition is that an herb is a nonwoody plant that dies down to its root each winter. But many medicinal plants are native to the tropics, where there is no winter and the plants do not seasonally die down to the ground. A far better term is actually "botanical medicine," or "botanicals" for short, but these terms are not familiar to most people. As a consequence, most people stick with the term "herb" to describe any medicinal plant. Basically, a medicinal herb is any plant that can benefit health.

Q. How common is herb use in the United States?

A. Sales of botanical products in the United States now exceed $3 billion annually, while sales of all pharmaceuticals amount to between $50 billion and $60 billion. In a 1998 study, Hartman and New Hope surveyed 43,442 American households and found that 68 percent had consumed a vitamin, mineral, or herbal supplement in the previous six months. According to a recent survey, 32 percent of adults spend an average of $54 per year on herbal remedies, for a total of $3.24 billion. Despite this, North America still lags far behind Europe in both herb use and research.

Q. Do herbs really work?

A. Yes! About 80 percent of the world's population uses herbal medicines, from the most basic folk remedies to the well-researched phytomedicines of Europe and Asia. Doctors who can choose whether to prescribe herbs or pharmaceuticals often make herbs their first choice. For example, in Germany, doctors prescribe St. John's wort twenty to one over the leading antidepressant drug because the herb is just as effective, costs less, and has none of the drug's side effects.

Q. What are some of the older systems of herb use?

A. Chinese and Ayurvedic (East Indian) medicine are two ancient systems that have not only withstood the test of time, but are now holding their own in regard to research findings. Both use the herbs of their own geographical areas, often in specific combinations, to treat a multitude of conditions. Both the Chinese and Ayurvedic systems are comprehensive, treating the body and mind as an interacting unit.

Q. These older approaches seem to take a very long time to work. Are they less efficient?

A. Unlike western medicine, the diagnoses and treatments are adapted to the individual—one pill does not fit every body. These ancient systems pay attention to all aspects of each individual. This takes time. In these approaches, treatment is designed specifically for you, aimed at building your health and not just fixing your current ailment. This really makes sense: from our faces to fingerprints, no two of us are alike.

Q. How do herbs work?

A. Herbs are rich in vitaminlike compounds such as polyphenols and flavonoids. They also contain many actual vitamins, minerals, and carotenoids. Believe it or not, many of these substances are also found in fruits and vegetables. In a sense, an herbal medicine is a concentrated source of polyphenols and flavonoids, which, like all nutrients, help your body function at its optimal level.

Q. Is there anything to the idea that herbs contain "energy"?

A. The systems of herbal medicine, such as traditional Chinese medicine, look beyond the chemical and molecular compounds of herbs to their energetic aspects. This may not sound scientific, but it does makes sense: Plants absorb the energy of the sun to do their work. When we eat these plants, we take in this transformed energy and use it to supply our own energy. The chemicals in our bodies are basically shells that carry energy. The Chinese call our vital energy force *Qi*. Illness is seen as a blockage of the flow of this vital force. Similarly, Native Americans believe that the spirit (in effect, the energy) of the herb is directly involved in the battle with illness.

Q. How is our "energy" or "Qi" related to herbs?

A. All living things, from plants to humans, have energy circulating throughout themselves. When we are ill, it may be the result of a block in our ability to access and utilize our full energy potential. The use of herbs can unblock us. An analogy is a dammed river. The stagnant water breeds mosquitoes. We can use toxic pesticides to kill the mosquitoes, but that would also hurt the fish—and us, if we drink the water. Or, we can release the dam (the block) and allow the free flow of the water, and the mosquitoes will leave Similarly, during illness, the energy flow in the body may be blocked. An herb can release the block, allowing the free flow of the body's energy and enabling the healing processes. Science tends to describe such processes in a difficult way, but whether you think like a scientist or not, you will find that herbal medicines do work.

Q. How do herbal medicines differ in action from pharmaceuticals?

A. Herbs generally work with your own body processes to fight off disease. For example, echinacea increases the activity of your infection-fighting white blood cells. Often, an herbal medicine, which is made from the whole plant, works better than its separate, isolated ingredients. Unlike Western medicine, which looks for the active ingredient of a plant or other substance, herbal medicine relies on the synergistic action of the whole plant, which may contain thousands of compounds. All these compounds promote health in concert, whether they're in the plant or a person. Herbs tend to work more gradually, helping to strengthen the body's defenses. Drugs, on the other hand, have a more rapid and targeted action, which is also why they are more likely to cause side effects.

Q. Aren't many drugs derived from plants?

A. About 25 percent of all drugs are made from, or are synthetic duplicates of, chemicals originally found in plants. Some examples are morphine, a potent painkiller, which comes from the opium poppy; digitalis, a cardiac stimulant, which comes from foxglove; and reserpine, a sedative and antihypertensive, which comes from rauwolfia (Indian snakeroot).

Q. Are herbs safe?

A. To answer this question, let's start with a statistic: There are as many as 100,000 deaths in hospitals a year from the use of prescribed drugs, with more than 2,000,000 serious reactions to prescription drugs. In contrast, deaths from herbal medicines are extremely rare. The most common negative reactions are allergic symptoms and upset tummies. There are also some people who react negatively to an herb or drug that most other people tolerate. I always warn my patients that if they feel any ill effects from an herb they are taking, they should immediately reduce the dose or stop taking it. This is good common sense. If you feel you are having a side effect, trust your senses and act accordingly.

Q. Can you mix herbs with prescription drugs?

A. In general, yes—but it's worthwhile checking with your doctor before treating the same condition with both a drug and an herb. I routinely prescribe certain combinations of drugs and herbs. For example, patients taking a drug metabolized by the liver, such as an antidepressant, should also take milk thistle, an herb that supports normal liver function. I also recommend echinacea to strengthen the immune system after taking a course of antibiotics. There are some bad combinations, too, such as St. John's wort and a monoamine oxidase (MAO)–inhibiting antidepressant, and ginkgo and an anticoagulant (blood thinner).

Q. What about taking herbs during pregnancy and while nursing?

A. During pregnancy and while nursing a baby, women have to be cautious about their herbal intake, since any herbs they take will also affect the baby. Many herbs have not been approved for this, and I refer to the guidelines in the authoritative German *Commission E Monographs*.

Q. What is the German *Commission E Monographs*?

A. The German *Commission E Monographs*, now available in English translation, is a collection of reports, based on safety and efficacy data, on over 200 herbs. The reports were written by a group of researchers and clinicians in Germany for that country's equivalent of the U.S. Food and Drug Administration (FDA).

Q. Can children take herbs?

A. Yes, and often herbs are the treatment of choice. It is important that you follow sound advice when adjusting the dosages of herbal medicines for children. An excellent book for help with this is *Smart Medicine for a Healthier Child: A Practical A-to-Z Reference to Natural and Conventional Treatments for Infants and Children*, by Janet Zand, Rachel Walton, and Bob Rountree (Garden City Park, NY: Avery Publishing Group, 1994). In general, the adult dosage should be adjusted according to the child's weight. The easiest form for a child to take is a liquid, especially one with a glycerin base, which is sweet, though alcohol tinctures are more potent. If you disguise them in applesauce or mashed potatoes, you may be able to sneak them in.

Q. Why is research necessary to prove that herbs work?

A. Research studies are the backbone of conventional medicine. They have provided us with a wealth of valuable information in all fields, including human biology and the health sciences. The whole issue, however, is complicated by nonscientific factors that influence which therapies are given funds for research and which are not. Research is costly, running into the millions, and most natural products do not have any major funding behind them. In Europe—where many herbal medicines are classified alongside pharmaceutical drugs, prescribed by doctors, and covered by the national health plans—it's a different story. Because herbal medicine is accepted in Europe as a legitimate form of therapy, the drug companies there have the financial incentive to do the necessary research.

Q. Why are drugs better researched than herbs in the United States?

A. In North America, the pharmaceutical companies bear the cost of research only if they can patent the new products that result from it. For the most part, natural products such as herbs are not easily patented, and discoveries made about them become public domain. Because companies don't have exclusivity, they're reluctant to invest in herbal research that might also aid their competitors.

As a result of the drug companies' huge investments in their products, and of their collaboration with medical schools for research purposes, much of the information taught to doctors in medical school and beyond comes from the companies themselves. However, demand is growing in North America for a more natural approach to medicine, including the use of herbs. Therefore, it is likely that the American pharmaceutical industry will follow the European lead and create refined extracts, which can be patented.

Unfortunately, this approach involves focusing on the so-called active ingredients and removing the extra materials that actually make herbs powerful, effective, and safe. Since existing herbal research is pointing in the direction of whole extracts, this approach may be continued in the United States. There is still a great deal being discussed on this and other topics. In the meantime, however, we will have to be careful that herbs do not become solely prescription items, thereby restricting their use.

Q. How can I work with my doctor when he seems to be so against herbal medicines?

A. In the best of all possible worlds, your doctor would be familiar with herbal remedies and would prescribe them as needed. I believe most doctors are motivated and curious to find the best, least harmful approaches to helping their patients. I therefore recommend that you take this book or something similar to your doctor to introduce him to the benefits of herbal medicine. He may be skeptical, but draw his attention to the scientific references at the back of the book and encourage him in a non-argumentative way to look them up and read them.

Sharing this knowledge can help you, your doctor, and his or her other patients.

Remember, there are times when it's important to seek professional medical help—for example, in cases of high blood pressure, liver ailment, enlarged prostate, severe depression, or deteriorationg mental function. All are potentially serious conditions and should be checked out before you embark on a self-treatment program.

2.

Echinacea

Echinacea, pronounced *e-ki-NAY-sha*, is a popular herb often taken at the first sign of a cold or the flu. I have relied on it for my patients, my family, and myself for years. In this chapter, we will look at what echinacea does and how it works, and at the research that backs it up. At the times when it doesn't prevent a cold, it still shortens the length and severity of the illness, as we shall see. Not just a quick fix, however, echinacea can be used over time to increase resistance to bacterial, fungal, and viral infections.

Q. What is echinacea?

A. *Echinacea purpurea*, or purple coneflower, is a decorative plant that has been one of the most popular herbal medications in both the United States and Europe for over a century. Echinacea is the primary remedy for minor respiratory infections in Germany, where doctors write more than 1.3 million prescriptions for it annually. Native Americans use a related species, *Echinacea angustifolia*, for a wide variety of problems, including respiratory infections, inflammation of the eyes, toothache, and snakebite. In the nineteenth century, before the advent of sulfa drugs, echinacea was the leading cold and flu remedy in the United States.

Q. What does echinacea do?

A. Echinacea is useful for preventing and treating colds and flus, ear infections, bronchitis, bladder infections, and even yeast infections. Unlike antibiotics, which are of little benefit in these types of infections, echinacea can greatly reduce the annoying and painful symptoms. It works best for those who are prone to getting colds, since it boosts their weak immune systems.

Q. How does echinacea work?

A. Studies have shown that echinacea can ease cold and flu symptoms in people. For the details of how it works, however, we have to rely on animal and test-tube studies.

Echinacea enhances the immune system, the complex combination of responses that fights invaders such as bacteria and viruses. The immune system fights infections by producing antibodies, which are specific molecules made by blood cells in response to a specific antigen, or invader. The next time these antibodies encounter these antigens, they recognize them and fight them off. Without the immune system, we would succumb to every infective agent we encounter.

Echinacea stimulates antibody production, raises the white-blood-cell count, and stimulates the activity of the white blood cells that fight infection. These white blood cells include lymphocytes, which fight viruses; natural killer cells, which attack tumor cells; and macrophages, which gobble up disease-causing bacteria.

Q. Has any good research been done proving that echinacea works?

A. There are over 400 published studies on echinacea, 11 of these were controlled and double-blind. This means that half the volunteers were given the real product, and half, placebo or dummy pills. Neither the person nor the reasercher know till after, which was given, so as not to influence the result. One of these followed 108 patients with acute flu-like illnesses. Half recieved echinacea, the other half placebo. After 8 weeks, the results showed that the treated group stayed healthier longer, and, if they did get sick, had a shorter, less severe illness. One of these followed 108 patients with acute flu-like illnesses. Half of the patients were given echinacea, and the other half were given a placebo. After eight weeks, the results showed that the treated group stayed healthier longer, and the people from that group who did get sick had shorter, less severe illnesses.

Another double-blind study of echinacea and its effect on flu-like illnesses followed 180 people who were divided into three groups. One group was given a placebo, another group 450 mg of echinacea, and the third group 900 mg of echinacea daily. By the third day, the subjects receiving the higher dose of echinacea showed significant reductions in their flu symptoms, which included chills, sweating, sore throat, muscle aches, and headache. There was no improvement in the other two groups. This indicates the effectiveness of the herb, as well as the importance of taking an adequate dose.

Q. Can echinacea really treat yeast infections?

A. Hard-to-treat, recurring yeast infections often yield to echinacea. In one study, researchers gave 203 women with chronic vaginal yeast infections either oral doses of echinacea or a topical antifungal cream medication. The herbal group did 3.5 times better than the medicated group. Only 16.7 percent of the echinacea group had a recurrence of the infection, as opposed to 60.5 percent of the medicated group. The echinacea also boosted the women's overall immunity. Since recurrent yeast infections can be bothersome, to say the least, and do not always respond to conventional treatment, this is a significant finding. By supporting the immune system rather than simply treating the symptom, the patient does much better.

Q. Can children take echinacea?

A. Yes. As we all know, once children start day care or school, they often pick up colds, flus, and ear infections from exposure to other kids. An excellent preventive measure during the cold and flu season is a daily dose of echinacea. Children prefer a glycerin-based tincture or a tea, which can be given two to three times a day. A two-day break is recommended every eight weeks. One of my patients brought in her five-year-old son, who had been catching every bug that came to school, then passed it on to the rest of the family. I suggested the glycerin-based tincture for him and the alcohol-based tincture for his mother. Both remained free of illness for the rest of the season!

Q. What is the recommended dosage of echinacea?

A. The typical daily dosage of powdered echinacea extract is 300 mg three times a day. Alcohol tinctures of one part echinacea to five parts alcohol are usually taken at a dose of 3 to 4 ml three times daily; echinacea juice at a dose of 2 to 3 ml three times daily; and whole dried root at 1 to 2 gm three times daily. Echinacea is usually taken at the first sign of a cold and continued for seven to fourteen days. It can also be taken continuously for prevention, with two days off every two months. This break is recommended so that the body does not become too accustomed to the echinacea, causing the echinacea to suffer a reduction in efficacy.

Q. Is any one form of echinacea better than another?

A. Many herbalists feel that the liquid forms of echinacea are more effective than the tablets or capsules. Part of echinacea's benefit may be due to direct contact with the tonsils. The liquid forms may also contain more of the active ingredients, or possibly the ingredients are in a form that is more easily absorbed.

Unlike the healing powers of manufactured drugs, those of herbs are likely due to more than just the molecules involved. The energy field of the plant meeting the energy field of the individual may create an additional healing power. We can begin to understand this concept by looking at the success of acupuncture, which is based on unblocking the energy flow.

Echinacea is frequently combined with goldenseal (*Hydrastis canadensis*) in cold preparations. A prime ingredient of goldenseal, berberine, appears to activate the white blood cells.

Q. How safe is echinacea?

A. Echinacea is very safe, even in extremely high doses. Side effects are rare and are usually limited to minor gastrointestinal symptoms, increased urination, or a mild allergic reaction.

Germany's Commission E warns against using echinacea in cases of autoimmune disorder, such as multiple sclerosis (MS), lupus, or tuberculosis. These warnings are based on fears that echinacea might actually overstimulate immunity. There are also warnings that echinacea should not be used by persons with acquired immune deficiency syndrome (AIDS). Low doses can be used in the early stages to treat any accompanying infections, but not in more advanced AIDS. Echinacea is also safe for pregnant women. There are no known drug interactions.

3.

Garlic

The human cultivation of garlic (*Allium sativum*) goes back at least 5,000 years, and today this herbal medicine can be found almost everywhere in the world, from Polynesia to Siberia. In the first century A.D., Dioscorides, Hippocrates, and other ancient Greek physicians recommended garlic for many conditions, including respiratory problems, parasites, and poor digestion. Garlic is principally used to prevent and treat heart disease, hardening of the arteries, high blood pressure, and high levels of cholesterol and triglycerides.

Q. Does garlic really help treat heart disease?

A. Garlic is used to help prevent and reverse atherosclerosis, the hardening of the arteries that underlies high blood pressure, heart disease, and stroke. The evidence comes from both human and animal studies. Garlic reduced the size of plaque deposits, the "hard" material in hardening of the arteries, by nearly 50 percent in humans, rats, and rabbits. In a recent study of 200 men and women over a two-year period, those who took 300 mg or more of garlic daily showed improvement in the flexibility of their aorta, the main artery carrying blood from the heart. Garlic extracts have been shown to reduce blood pressure in dogs and rats, and numerous animal studies have shown that it can reduce blood clotting. This all makes garlic a remarkably effective treatment for arterial disease.

Q. Can garlic reduce cholesterol levels?

A. High blood levels of cholesterol and triglycerides, which are lipids (fats), are related to a higher incidence of heart disease. Thus, physicians recommend keeping your cholesterol and triglyceride levels down. Especially important to control is your low density lipoprotein (LDL), or "bad" cholesterol. The high density lipoprotein (HDL) form, or "good" cholesterol, is protective, while the LDL form is destructive to the arteries.

At least twenty-eight controlled clinical studies have shown that garlic can lower total cholesterol levels by about 9 to 12 percent, as well as improve the ratio of good to bad cholesterol. In a 1990 German study, 261 patients were given either 800 mg of standardized garlic or a placebo daily. Over the course of sixteen weeks, patients in the garlic group had a 12-percent drop in their total cholesterol and a 17-percent decrease in their triglyceride levels.

A European study comparing garlic to the drug bezafibrate found garlic to be just as effective at lowering cholesterol, and without the drug's side effects. Like prescription drugs, garlic appears to interfere with the manufacture of cholesterol in the body. Some studies have been less successful. The differences may be due to manufacturing differences in the particular garlic used. I'll explain more about this shortly.

Q. Does garlic act as a natural antibiotic?

A. Many studies, some dating back to the 1940s, have shown that garlic has powerful antibiotic and antiviral properties. Persons with human immunodeficiency virus (HIV) have taken garlic to prevent secondary bacterial and yeast infections, a use based on scientific research. Garlic has been used to treat certain viral and fungal infections, including intestinal, oral, and vaginal candida. Albert Schweitzer, MD, used garlic to treat amebic dysentery in Africa. These effects are best with raw garlic, or a substance with an allicin content equivalent to raw garlic.

Q. What is allicin?

A. The primary active ingredient in garlic is allicin, a sulfur-containing compound that the body converts to other therapeutic compounds. Allicin is found only in garlic products produced by crushing the fresh bulb, not in those produced by the steam distillation of the oil.

Q. Is it true that garlic can prevent cancer?

A. Several large studies strongly suggest that diets high in garlic can prevent cancer, including colon, esophageal, and stomach cancers. In one study, a group of 41,837 women were questioned in 1986 about their lifestyle habits. Four years later, follow-up questionnaires showed that women whose diet includes a significant amount of garlic are approximately 30-percent less likely to develop colon cancer. This is a good reason to eat your garlic!

Q. What is the recommended dosage of garlic?

A. One or two raw garlic cloves a day should be sufficient, but many people prefer to avoid the social after-effects of garlic. Some companies sell odorless substitutes made from the allicin in garlic. There are also enteric-coated garlic-powder supplements. (The enteric coating delays digestion of the tablet until it passes from the stomach to the intestines.) For health maintenance, 2,500 mcg of allicin daily is enough, while for therapeutic purposes, at least 5,000 mcg of allicin a day is necessary. You could also encourage your family and friends to eat garlic. This way, no one will notice your breath!

Q. Does garlic have any cautions or side effects?

A. The only common side effect that results from taking garlic is the characteristic unpleasant breath odor. Even taking an odorless supplement offers no protection, since this form produces an offensive smell in up to 50 percent of users. Other side effects, such as nausea, headache, sweating, and dizziness, occur only rarely.

When raw garlic is taken in excessive doses, it can cause numerous symptoms, such as heartburn, nausea, stomach upset, vomiting, diarrhea, flatulence, facial flushing, rapid pulse, and insomnia. When applied to the skin, garlic can cause skin irritation, blistering, and even burns.

Since garlic thins the blood, it would be prudent to avoid taking high-potency garlic pills prior to surgery or if you are already taking prescription blood thinners. However, garlic is presumed to be safe for pregnant and nursing women. Nursing babies actually seem to like the taste.

4.

Ginkgo

Ginkgo biloba, or ginkgo, as it's commonly known, is the most widely prescribed herb in Germany. More than 6 million prescriptions are written there for ginkgo in a typical year. Used mainly to treat failing mental faculties, including memory loss, in the elderly, it is also used for a variety of circulatory problems. Over 200 million years old, the ginkgo is the oldest surviving species of tree on the planet, and individual trees may live for 1,000 years. The bi-lobed—that is, double-lobed-leaf gives the plant the name "biloba." Since the 1950s, the focus of medical research has been on the extracts of ginkgo leaves.

Q. For what do most people use ginkgo?

A. Ginkgo is mainly prescribed for elderly people with impaired mental function ("dementia"), whether due to Alzheimer's disease or poor blood circulation in the brain. An article in the respected medical journal *Lancet* concluded that the results of more than forty double-blind controlled studies proved that ginkgo extract is indeed an effective treatment for this.

Recently, the *Journal of the American Medical Association* reported on the results of a year-long double-blind trial of ginkgo's effect on mental function involving more than 300 patients. The participants were given either 40 mg of ginkgo extract or a placebo three times daily. The patients taking the ginkgo did better than those receiving the placebo.

Q. What is Alzheimer's disease, and how does ginkgo help it?

A. A serious and growing problem, Alzheimer's disease, or "senile dementia", literally means "impaired mental function of the elderly." It affects approximately 4 million Americans, which includes nearly 30 percent of the people over the age of eighty-five.

Researchers have found that ginkgo improves memory and the ability to concentrate, elevates the mood, and relieves dizziness and anxiety. Moreover, taking ginkgo actually stops or significantly slows down the progression of Alzheimer's. This can enhance the quality of life and improve the ability to function adequately. It can take up to twelve weeks of treatment for ginkgo to take full effect, so don't give up too quickly. However, be aware that while ginkgo can reduce the symptoms and progression of Alzheimer's in many people, it will not cure the disease.

Q. Can ginkgo help with depression?

A. Ginkgo also appears to improve depression. This may be due not only to the increase in circulation, but to ginkgo's direct effect on the "feel-good" chemical messengers, known as neurotransmitters, in the brain.

Q. Can ginkgo be used successfully in younger people?

A. Several studies have shown that ginkgo helps brain function in younger people as well. It works in part by increasing the blood flow in the brain—that is, by improving how well the brain's cells are nourished. Students who use ginkgo in the morning just before going to school to take an exam find a noticeable improvement in their ability to recall information. It appears to work best for this purpose when taken as a loading dose just before it's needed than when taken long-term. Not only does ginkgo improve young peoples' short-term memory and concentration, but brainwave (EEG) tests have also shown that it enhances their mental function.

Q. What about using ginkgo for poor circulation elsewhere in the body?

A. Germany's *Commission E Monograph*s recommend ginkgo for the treatment of poor circulation in the legs due to hardening of the arteries. This circulatory disorder can cause a painful condition called intermittent claudication, which interferes with walking, thus further affecting circulation. After about six weeks of treatment with ginkgo, there is often a marked improvement in this hard-to-treat condition. According to the *Commission E Monographs*, at least four double-blind studies have shown that ginkgo can increase the pain-free walking distance by about 75 to 500 feet.

Q. What is the scientific evidence for this?

A. Studies have shown that ginkgo extracts can improve circulation in general. We don't know exactly how it does this, but it appears to make the blood more fluid, reduce the tendency toward blood clots, and extend the life of a natural blood-vessel-relaxing substance, further enhancing blood flow.

Q. Can ginkgo help with sexual problems?

A. Male impotence is often due to poor circulation in the pelvic area. Since ginkgo increases the general circulation, it often helps to restore sexual function. One European study showed that ginkgo supplements can correct erectile dysfunction, but it may take several months for this effect to kick in.

Q. For what else is ginkgo used?

A. Ginkgo is also used to treat water retention in women with premenstrual syndrome (PMS). Another use is for tinnitus, a chronic, uncomfortable ringing in the ears. Ginkgo helps tinnitus most likely because of its ability to increase the circulation within the ear. A powerful and protective antioxidant, ginkgo may also be useful in treating other age-related conditions, such as macular degeneration and the complications of diabetes.

Q. What makes ginkgo an antioxidant?

A. Our bodies contain free radicals both as a result of our metabolism and ingested with our air, water, and food. Free radicals cause damage to our cells, and their effect increases with age. In fact, one cause of aging is believed to be free-radical damage. Antioxidants bind to free radicals, making them harmless, and are important for maintaining health and youth. Ginkgo's active compounds are antioxidants, and this is likely one mechanism by which the herb protects the brain, the blood vessels, and the circulatory system in general. I'll explain more about antioxidants in reference to milk thistle in chapter 7.

Q. What exactly are ginkgo's active ingredients?

A. Ginkgo's effect has been attributed to compounds called flavone glycosides and ginkgolides. Nearly all the studies done with ginkgo have used a leaf extract standardized to 24-percent flavone glycosides and 6-percent ginkolides. These substances are somehat similar to the flavonoids found in fruits and vegetables.

Q. What is the recommended dose?

A. The typical dose of ginkgo is 40 to 80 mg three times daily of an extract of fifty parts alcohol to one part ginkgo standardized to contain 24-percent ginkgo-flavone glycosides. You may need to take supplements for six weeks before noticing results, so don't give up too soon. For some conditions, such as peripheral vascular disease, tinnitus, and dizziness, the recommended dose is higher—80 mg twice daily or 40 mg four times daily.

Q. How safe is ginkgo, and does it have any side effects?

A. Ginkgo is very safe. Extremely high doses have been given to animals without serious consequences. Ginkgo has shown no toxicity to the liver or kidneys, and has not hampered new blood-cell formation. We still have no proof of its safety for pregnant and nursing women, however.

In all the clinical trials using ginkgo, involving a total of almost 10,000 patients, the incidence of side effects produced by ginkgo extract was extremely small, with just a few cases of nausea, heartburn, mild tension headache, dizziness, and allergic skin reactions. However, massive ginkgo overdoses have led to agitation, restlessness, and gastrointestinal distress.

According to the *Commission E Monographs*, German medical authorities do not believe that ginkgo interacts seriously with any drugs. However, because of ginkgo's blood-thinning effects, some authorities caution that it should not be combined with anticoagulants or even aspirin. There have been two case reports in highly regarded journals of sub-dural hematoma (bleeding in the skull) and hyphe-ma (spontaneous bleeding into the iris chamber) in association with ginkgo use. Yet, if this risk is signif-icant, it seems odd that similar side effects were not observed in the large number of patients who partic-ipated in the ginkgo clinical trials or have been using the herb for years in Germany.

5.

Ginseng

Ginseng is one of the most venerable herbs. It has been in continuous use in China, to restore vital energy, for over 2,000 years. There are actually three different herbs commonly called ginseng—Asian ginseng (*Panax ginseng*), American ginseng (*Panax quinquefolius*), and Siberian ginseng (*Eleutherococcus senticosus*). This last herb is actually not ginseng at all, although it is believed to function in a similar way. Sometimes called adaptogens, the ginsengs increase resistance to stress, enhance mental alertness, and improve stamina and immunity.

Q. How do the ginsengs differ from each other?

A. Asian ginseng is a perennial that grows in northern China, Korea, and Russia. Its close relative, American ginseng, is cultivated in the United States. In traditional Chinese herbology, Asian ginseng is used to strengthen the digestive system and lungs, calm the spirit, and increase the overall energy. The active ingredients in both Asian and American ginseng are substances called ginsenosides.

Siberian ginseng, also called eleuthero, grows mostly in China and Siberia. In the 1940s, the Russian scientist who first researched this less expensive herb concluded that it was as good as ginseng. It can be taken for a longer time than Asian ginseng, since it is less stimulating and has a milder effect on the hormones estrogen and testosterone.

Q. What is an adaptogen?

A. An adaptogen is an herb that helps the body adapt to stresses of various kinds, such as heat, cold, exertion, trauma, sleep deprivation, toxic exposure, radiation, infection, and psychological stress. It causes no side effects, is effective at treating a wide variety of illnesses, and helps return the body to homeostasis (balance). For example, if your blood pressure is too high, an adaptogen will lower it, and if your blood pressure is too low, an adaptogen will raise it, thus moving it toward normal. Because ginseng has all these characteristics, it is considered an adaptogen. It is often mistakenly called a stimulant, but it is actually more of a supportive tonic that works over time.

Q. What is ginseng used for today?

A. Ginseng appears to protect us from stress, which is a significant health problem in modern-day life. Among its adaptogenic effects, ginseng also stimulates the mind, increases physical performance, strengthens immunity, and helps the hormones to better regulate bodily functions. It helps to protect the liver, which might account for its ability to speed the processing of alcohol in the body. Ginseng also increases oxygenation in the cells and tissues of the body, thereby boosting endurance, alertness, and visual-motor coordination. Its effect on brain function makes it useful for the elderly, and it combines well with ginkgo for maximum effect.

Q. How does ginseng help a person deal with stress?

A. The ginsengs, especially Siberian, support the adrenal glands, our stress-management glands. The adrenal glands provide the energy for the fight-or-flight response. When we are overly stressed, the adrenals work overtime to release the emergency hormones adrenalin and cortisol. While necessary for true emergencies, such as running from danger, this overstimulation can wear us out over time. This leads to fatigue and poor immunity, making us more vulnerable to colds, flus, and even more serious illnesses. It is estimated that 60 percent of patient visits are due to stress-related illnesses, including heart disease, high blood pressure, and cancer. Thus, stress-reduction must be a priority in our self-care.

Q. What about ginseng's use for endurance?

A. Siberian ginseng is used by the Russian Olympic teams, especially the runners and weight lifters. Mountain climbers, sailors, and factory workers use it to enhance their performance and reduce the number of sick days they must take. In one experiment, radio operators who took 60 drops of ginseng daily for one month significantly increased their work capacity. In another study, skiers who took 3 droppersful before a race increased both their endurance and resistance to the effects of the cold. Thus, ginseng can be used both for athletic purposes and for enhancing work capacity.

Q. Does ginseng play a role in cancer treatment?

A. Yes, ginseng helps in cancer treatment in several ways. It protects the normal cells from radiation damage and eases the side effects of radiation treatment. It also stimulates the white blood cells that fight cancer. One study followed eighty patients with breast cancer. Half the group was given Siberian ginseng and showed a lower incidence of nausea, dizziness, and loss of appetite. However, ginseng is not by itself a cure or treatment for cancer.

Q. What other uses does ginseng have?

A. Ginseng is a useful adjunct in treating diabetes, helping to lower blood sugar both directly and by enhancing the production of insulin by the pancreas. It also acts to decrease cholesterol, while raising the HDL level. This makes it helpful in treating heart disease and high blood pressure.

Q. What forms does ginseng come in, and what is the recommended dose?

A. Ginseng is available as a powdered root in capsules or formed into tablets, or as an alcohol-based tincture. In general, it is better to take Asian ginseng than American ginseng. The recommended dose of Asian ginseng is 100 to 200 mg daily of an extract standardized to contain 4- to 7-percent ginseno-sides. Siberian ginseng should be taken at a dose of 200 to 400 mg daily of standardized extract containing more than 1-percent eleutherosides. The recommended dose for ginseng tincture is 5 ml twice daily of a concentration made of five parts alcohol to one part ginseng. Take ginseng for three weeks, then take one week off.

Q. Does ginseng have any side effects?

A. Side effects are rare. However, allergy can occur, as it can with any substance. With Asian ginseng, menstrual abnormalities and breast tenderness have been reported. Overuse can cause over-stimulation, including insomnia. There have also been unconfirmed reports of excessive doses raising the blood pressure and increasing the heart rate. In Chinese traditional medicine, ginseng is used during pregnancy, but as with any herb, this should be done only under the care of a qualitified health-care practitioner.

6.

Kava

Kava (*Piper methysticum*) is a member of the pepper family that has been cultivated by Pacific Islanders for over 3,000 years for use as a social and ceremonial drink. The first description of kava came to the West with Captain James Cook, who traveled through the South Seas in a number of celebrated voyages. To this day, when village elders or others in the Pacific Islands come together for a significant meeting, they begin with an elaborate kava ceremony. Currently, kava is used in Europe and increasingly in the United States to treat stress, anxiety, and insomnia.

Q. How is kava used today?

A. According to Germany's *Commission E Monographs*, as well as my own clinical experience, kava is useful for relieving states of nervous anxiety, stress, and restlessness. It works both on the limbic system, the emotional center of the brain, and on the muscles, thus promoting relaxation in two different ways. Compared to prescription drugs, kava, taken in the lower doses, does not impair mental function and is not addictive. Its muscle-relaxing effects make it particularly useful for treating headaches, backaches, and other tension-related pain.

Q. How is kava used for stress?

A. I highly recommend kava for handling stressful situations. I used it when writing a book under a tight deadline, and found it really helped. A 60 to 70 mg dose of kava is excellent for such challenging events as giving a speech (most people's worst fear!), taking an exam, and traveling by airplane if you have a fear of flying. A dose of kava definitely takes the edge off, without affecting mental sharpness.

The *Commission E Monographs* recommend using kava for no more than three months without medical supervision. This caution is more to encourage the pursuit of other therapeutic approaches such as psychotherapy or stress-reduction techniques than for reasons of safety.

Q. What about kava's use for sleep?

A. Kava is an excellent bedtime sedative, to be taken in place of prescription drugs. It has the advantage of not suppressing rapid eye movement (REM) sleep, the stage in which dreaming and inner mental processing take place. People report waking up refreshed and ready to start the day. The commonly used sedatives—the benzodiazepines such as Restoril, Ativan, and Valium—suppress REM sleep and often lead to a groggy, hungover feeling the next morning. Moreover, these prescription drugs are addictive, causing a withdrawal syndrome that includes rebound insomnia, which is often worse than the original problem.

Q. While kava calms your mind, does it also impair your mental abilities?

A. Fortunately, unlike the benzodiazepines, kava does not seem to impair mental functioning, and might even improve it. An herbalist colleague, Terry Willard, even prescribes the herb for professional hockey players, who report that it enhances their concentration on the ice. Research showed that kava actually enhanced reaction time and performance on a word-recognition test.

Q. What is the scientific research on kava?

A. The largest and longest-running study on kava to date and published in 1997 was conducted by a

German researcher, H.P. Volz. He evaluated just over 100 patients with anxiety-related disorders using the Hamilton Anxiety Scale (HAM-A), which measures the symptoms of anxiety, such as restlessness, nervousness, stomach discomfort, heart palpitations, chest pain, and dizziness. Over a six-month period, he gave half of the patients kava in the form of 300 mg a day of a 70-percent kavalactone extract and half a placebo. The kava group showed a significant improvement in their symptoms over the control group. More-over, the kava group actually suffered fewer side effects than did the placebo group.

An earlier study had had similar success. It followed 58 patients with anxiety, half of whom were given 300 mg a day of a 70-percent kavalactone extract and the other half of whom took a placebo. Again, the people who took the kava showed a significantly greater improvement in their symptoms.

Q. What about research comparing kava to tranquilizers?

A. A double-blind study tracked 174 patients with anxiety symptoms for a period of six weeks. The patients were given either 300 mg a day of a 70-percent kavalactone extract, 15 mg a day of oxazepam, or 9 mg a day of bromazepam. Oxazepam and bromazepam are both benzodiadepines. All the groups did about the same, although the medicated groups suffered more side effects.

Unlike the benzodiadepines, kava does not seem to impair driving ability. In one test, in fact, it actually enhanced driving ability. However, at the higher doses, kava can be sedating, so appropriate caution is needed. You should not drive after taking a significant dose of kava.

Q. Is kava useful for treating menopausal symptoms?

A. Clinical experience and research, including an eight-week study, show kava to be an excellent treatment for menopausal symptoms, including anxiety and hot flashes. In treating menopausal symptoms, kava is often combined with other herbs, such as black cohosh and dong quai, and other supplements as well.

Q. What are the active ingredients in kava, and how do they work?

A. By 1966, researchers had isolated a combination of ingredients in kava, called kavalactones. The major ones are dihydrokavain, kavain, methysticin, and dihydromethysticin, which produce sedative, painkilling, and anticonvulsant effects. High doses of kava extract cause muscular relaxation. Kava is also a local anesthetic. In fact, if you drink kava extract, you will experience a numb feeling in your mouth similar to the Novocain your dentist gives you.

Q. What forms does kava come in?

A. Kava comes as tablets and various types of capsules, such as hard gelatin, vegetable, and soft-gelatin capsules. It also comes in tincture form, if you can tolerate the bitter taste. An acquired taste, kava is actually enjoyed by many people. The tincture form of kava is absorbed more rapidly, as is the spray form. You can also buy powdered kava, usually in a South Sea Island store, and extract it yourself, native style!

Q. What is the recommended dosage of kava?

A. Kava is available in a standardized form, in which the total amount of kavalactones per capsule or tablet is listed. For use as an anti-anxiety agent, use enough kava to supply about 40 to 70 mg of kavalactones. Take three doses a day, but do not take more than a total of 300 mg of kavalactones a day. While you can feel kava's effects immediately, the full anti-anxiety effect often takes about one week to develop and about four to eight weeks to reach its full potential. The usual dose of kava for insomnia is 210 mg of kavalactones one hour before bedtime.

The label on the bottle will likely indicate that the kava is a standardized extract with 30-percent kavalactones. This is in contrast to the 70-percent extract cited in most of the research on kava. All you need to do is adjust the number of pills to obtain the proper amount of kavalactones. You will see kava supplements containing anywhere from 200 to 250 mg of 30-percent extract, which translates into 60 to 75 mg of kavalactones. The equivalent of this in a 70-percent extract is 100 mg, yielding 70 mg of kavalactones. Confused? Just go by the milligrams of kavalactones per dose, not the total weight. In the ostensibly weaker extract, more of kava's other components are present. We know from traditional use and from research on kava and other herbs that the whole herb is generally best, offering a synergistic effect arising from the natural combination of the ingredients.

Q. Does kava have any side effects?

A. There are occasional reports of headaches, gastrointestinal distress, and allergic rashes resulting from the use of kava. Long-term use (months to years) of native-prepared kava at doses in excess of 400 mg of kavalactones per day can create a characteristic generalized dry, scaly skin condition. This side effect does not occur with the normal doses of commercially prepared products. In any case, the rash does go away when the kava use is stopped.

Addiction is not a problem with kava, nor is withdrawal. However, high doses of kava can cause intoxication, so there is some concern that it could become an herb of abuse. There have been reports of young people trying to get high by taking products they thought contained kava. One of these products turned out to contain dangerous drugs, but no kava at all.

Kava should be used with caution when driving. Also, the German *Commission E Monographs* warn against the use of kava during pregnancy and while nursing.

Q. Does combining kava with a drug cause any problems?

A. Although there are no proven drug interactions, the *Commission E Monographs* caution that kava should not be combined with alcohol or with prescription tranquilizers or sedatives, or other depressant drugs, except under a doctor's supervision.

There is a case report in the medical literature of

an individual who took kava with the benzodiazepine alprazolam (Xanax) and ended up in the hospital drowsy and disoriented. He recovered within hours, but the warning is there: Mixing two different tranquilizers, an herb and a drug in this case, is unsafe. For this same reason, patients should not attempt to make a transition from a benzodiazepine to kava without medical supervision, since you may need to be completely off the drug before you can start taking the kava.

Q. What about combining kava with other herbs?

A. There are many herbs that go well with kava. For example, I often prescribe kava in combination with St. John's wort for patients who are both anxious and depressed. The rapid activity of kava is particularly helpful in these cases, since St. John's wort generally takes longer to work. While the safety of this combination has not been officially established, kava and St. John's wort are used together a great deal in Gemany and increasingly in the United States. No adverse effects have been reported. (For a further discussion on combining these two herbs, see Chapter 8.)

Similarly, kava combines well with such other herbal tranquilizers as valerian, hops, and passionflower. One recent study using a product combining these herbs looked at kava's ability to help the test subjects better handle the stresses and hassles of everyday life. The results of the eight-week study were impressive. While the placebo group showed no improvement at all, the kava group scored much better both on the objective tests and the subjective questionnaires. In addition, most of the participants elected to continue taking the product after the conclusion of the study.

Q. How do you transition from taking a tranquilizer to taking kava?

A. Getting off a benzodiazepine such as Valium, Xanax, or Klonopin can be very difficult. Since the withdrawal symptoms can be severe and even life-threatening, the transition must be done very slowly and under medical supervision. Switching from an antidepressant or the tranquilizer Buspar is easier, but still requires a doctor's supervision.

7.

Milk Thistle

The seeds, fruit, and leaves of milk thistle (*Silybum marianum*) have been used for medicinal purposes for more than 2,000 years. The Roman writer Pliny the Elder, who lived from A.D. 23 to 79, reported that the juice of milk thistle mixed with honey "could carry off bile." In Europe, the herb was widely used up through the early twentieth century for the treatment of liver ailments, as well as insufficient lactation.

Q. What is milk thistle used for today?

A. The main constituent of milk thistle seeds, silymarin, is used to treat liver disease. Silymarin protects the cells against toxins and stimulates new cell growth in the liver. Based on the extensive folk use of milk thistle in cases of jaundice, European medical researchers have done serious research on the herb's medicinal effects. Milk thistle is widely used to treat alcoholic hepatitis, alcoholic fatty liver, liver cirrhosis, liver poisoning, and viral hepatitis. Milk thistle is one of the few herbs that has no equivalent in the drug world, with only two other natural substances, alpha-lipoic acid and N-acetylcysteine, having similar effects.

Treatment with milk thistle often produces an improvement in the symptoms of chronic liver disease, which include nausea, weakness, loss of appetite, fatigue, and pain. The blood levels of the liver enzymes, which are elevated in liver disease or damage, frequently go down.

Q. What are the active ingredients in milk thistle?

A. The active ingredients in milk thistle appear to be four substances known collectively as silymarin, of which the most potent is silibinin. When injected intravenously, silibinin is one of the few known antidotes to poisoning by the deathcap mushroom, *Amanita phalloides*.

Q. How does milk thistle work?

A. Silymarin is a powerful antioxidant. We are constantly exposed to toxins such as cigarette smoke, car exhaust, pesticides, and other chemicals in our air, food, and water. This is in addition to the toxins that our bodies produce as by-products of our own metabolism. All these toxins produce free radicals, which cause cell damage. They can, however, be neutralized by substances called antioxidants. Two major antioxidants produced by the body, glutathione and superoxide dismutase (SOD), are greatly enhanced by silymarin. Thus, milk thistle acts as an antioxidant in the liver, protecting it from free-radical damage. Animal studies suggest that milk thistle extract can also protect against many poisons, from toluene, a common solvent, to acetaminophen, the main ingredient in Tylenol.

In Europe, doctors often prescribe milk thistle as extra protection for patients taking medications that are known to cause liver problems. I often recommend it to patients who are on medications such as antidepressants, which are metabolized (broken down) in the liver. Milk thistle can also protect against future toxic exposure.

Q. How well does milk thistle work in alcoholic liver disease and liver cirrhosis?

A. Alcohol consumption and alcoholism take a heavy toll on the liver, which is the chemical factory of the body. It is here that alcohol is broken down into its metabolites. Silymarin is useful for preserving the liver, although research shows that abstinence is the best treatment. One study observed 106 Finnish soldiers with mild alcoholic liver disease. The group treated with milk thistle had a significant improvement in liver function as measured by blood tests and a biopsy. (In a biopsy, a small piece of liver tissue is examined under a microscope.) Other studies have shown similar results. Again, however, research shows that abstinence from alcohol is still a better treatment than milk thistle.

Two different long-term controlled studies showed how milk thistle prolonged the life of patients with liver cirrhosis. In one study, 170 patients were given either milk thistle or a placebo, and were observed for three to six years. After four years, 58 percent of the milk-thistle group had survived, compared to only 38 percent of the placebo group. Double-blind studies of patients with chronic viral hepatitis have shown that milk thistle can produce marked improvement in symptoms such as fatigue, reduced appetite, and abdominal discomfort, as well as in the levels of the liver enzymes.

Q. What about milk thistle's use in various forms of hepatitis?

A. Conventional treatments are not very successful against hepatitis, while silymarin, especially when combined with other beneficial nutrients, is an excellent treatment. A patient of mine, a fifty-year-old man with elevated liver enzymes due to an old case of hepatitis, responded well to milk-thistle treatment. His enzymes came down to normal after eight weeks, and his symptoms of depression, fatigue, and nausea cleared as well.

Q. What is the recommended dosage of milk thistle?

A. The standard dosage of milk thistle is 200 mg two to three times a day of an extract standardized to contain 70-percent silymarin complex. There is some evidence that silymarin bound to the nutrient phosphatidylcholine is better absorbed. This combination supplement should be taken at a dose of 100 to 200 mg twice a day. Medical supervision is essential in all cases of liver disease, since liver desease is a very serious condition.

Q. How safe is milk thistle?

A. Milk thistle and its silymarin extract are basically nontoxic, causing only the mildest of side effects in a small minority of patients. A study involving 2,637 patients showed a low incidence of side effects, limited primarily to mild gastrointestinal distress. Milk thistle is safe for use by pregnant and nursing mothers. Researchers have even felt safe enough to enroll pregnant women in the studies on silymarin.

8.

St. John's Wort

St. John's wort (*Hypericum perforatum*) is becoming the top natural treatment for mild to moderate depression, with all the benefits of prescription antidepressants, without the side effects, and at one-tenth the cost. This chapter discusses the basics of St. John's wort—what is it, how it works, what to expect, and how to use it.

Q. What is St. John's wort?

A. St. John's wort is a bushy perennial plant with yellow flowers that commonly grows wild. It is native to many parts of the world, including Europe, Asia, and the United States. It gets its unusual name from St. John the Baptist, since it was traditionally collected on St. John's Day, June 24. "Wort" is the Old English word for "plant."

Q. For what is St. John's wort currently used?

A. Research in treating patients with depression has shown that St. John's wort relieves the symptoms of sadness, helplessness, hopelessness, anxiety, headache, and exhaustion, and it does so with minimal side effects. St. John's wort is also useful in seasonal affective disorder and PMS, and as an antiviral and anti-inflammatory agent. Hypericum oil is useful for the treatment wounds, bruises, muscle aches, and first-degree burns.

Q. What has research shown about St. John's wort?

A. In 1996, a significant and extensive review article on St. John's wort was published in the respected *British Medical Journal*. The authors did a meta-analysis, or summary and comparison, of twenty-three randomized clinical trials to look for overall conclusions. Fifteen studies compared the herb with a placebo, and eight compared it with conventional antidepressants, in a total of 1,757 outpatients. The research showed that St. John's wort worked nearly three times better than the placebos, with a success rate of 55 percent, versus only 22 percent in the placebo groups.

Q. How does St. John's wort compare with antidepressant drugs?

A. The meta-analysis just mentioned also compared the herb with a number of antidepressant drugs, and showed that St. John's wort worked slightly better, eliciting a positive response 64 percent of the time versus 59 percent for the antidepressants. In addition, only 0.8 percent of the people taking the herb dropped out of the study because of side effects, whereas 3 percent of those on the drug treatment dropped out.

Q. What about using St. John's wort for stress and anxiety?

A. While St. John's wort won't take stress away, it will help you deal with it better. The herb should be taken regularly, at the usual doses, and not just before stressful events, since it needs to build up in the system to be the most effective. You can also add kava, 70 mg or so of standardized extract (30-percent kavalactones), three times daily at times of increased stress.

Q. Can St. John's wort help sleep?

A. The herb works with the body's own sleep-promoting mechanism, the release of melatonin, to bring on restful sleep, without the hangover or addictive effects of prescription sedatives. One study showed that a dose of 90 drops a day over a three-week period significantly increased the night-time melatonin level. Since it can take a week or so for this effect to begin, St. John's wort is recommended mainly for recurring insomnia, not just the occasional restless night.

Q. What about using St. John's wort for PMS?

A. PMS is a common complaint that occurs in many women about seven days prior to the onset of their period, with moodiness, irritability, bloating, and fatigue. Many women report that their PMS and menstrual cramps, or menopausal symptoms, stopped after they began taking St. John's wort for depression. Some women begin taking it just before the onset of PMS. Others find that the herb needs time to build up in the system, so they take it all month long, with extra doses as needed.

Q. Can St. John's wort help with weight loss?

A. The evidence isn't really clear on this. However, because St. John's wort curbs anxiety and depression, it may also help with "nervous eating."

Q. How can one herb produce so many different benefits?

A. Like most medicinal plants, St. John's wort is a complex substance with over two dozen active ingredients, each with its own effects. These compounds also work together to accomplish more than any one component could do on its own. Rather than unwanted side effects, you may receive bonus healing effects. One ingredient, hypericin, while not the main antidepressant, is used as the marker for standardization of St. John's wort products. Recent research by Professor W.E. Muller at the University of Frankfurt suggests that hyperforin may be the main antidepressant component.

Q. How does St. John's wort treat depression?

A. St. John's wort probably affects the levels and activities of various neurotransmitters. It was initially thought to be an inhibitor of MAO, an enzyme that breaks down neurotransmitters. However, it more likely acts to increase the availablity of the antidepressant neurotransmitters, including serotonin, norepinephrine, and dopamine. St. John's wort therefore is similar in action to the various antidepressant drugs.

Q. I have seasonal affective disorder (SAD), or the "winter blues." Can St. John's wort help?

A. St. John's wort has been quite successful in the treatment of SAD. The lack of sunlight that occurs in autumn and winter triggers biochemical changes in the brain and leads to such symptoms as depression, impaired concentration, anxiety, marked decrease in energy and libido, and carbohydrate cravings. SAD is especially common in countries at the extreme northern and southern latitudes, where there are fewer sunlight hours during the winter months. Yet, when affected individuals get their required doses of sunlight, they feel energetic and ready to get on with their lives. In a study comparing St. John's wort to light therapy, the researchers concluded that St. John's wort is as effective as light therapy. This herb really does "bring light into dark places."

Q. When and how often should I take St. John's wort?

A. The standard dose of St. John's wort is 300 mg three times a day of an extract standardized to contain 0.3-percent hypericin. Alternately, some people take 450 mg twice a day, or 600 mg in the morning and 300 mg in the evening. If the herb bothers your stomach, take it with food. For some, St. John's wort can be stimulating and thus should be avoided close to bedtime. Others report that it helps them to fall asleep more easily.

Q. How quickly does St. John's wort work?

A. Many of my patients report positive effects almost immediately, with a sensation of "a weight being lifted," decreased anxiety, and an enhanced ability to concentrate. As with most antidepressants, though, it may take three or four weeks before you notice a significant effect.

Q. When does St. John's wort not work?

A. Anyone with the symptoms of depression should get a thorough medical examination to rule out other possible causes of the symptoms. Medical conditions such as thyroid disorders, anemia, hypoglycemia, chronic fatigue syndrome, and nutritional deficiencies can also cause depression. More severe depression may also be unresponsive and, consequently, require stronger medication.

Q. Does St. John's wort have any side effects or cautions?

A. St. John's wort is essentially free of side effects, causing just infrequent bouts of mild stomach discomfort, allergic rashes, tiredness, and restlessness. There have been no published reports of serious adverse effects or drug interactions. Animal studies involving enormous doses for twenty-six weeks did not show any serious toxicity, either. Excessive sensitivity to sunlight can occur in fair-skinned individuals taking higher doses, so proper precautions should be taken. Older reports suggested that St. John's wort works like the class of drugs known as MAO-inhibitors, leading to restrictions such as avoiding aged cheese, red wine, and decongestants. However, this concern is no longer considered valid. St. John's wort has not been approved for use during pregnancy or while nursing.

Q. How long should I take St. John's wort?

A. St. John's wort has been used safely, with no ill effects, by many depressed patients in Europe for years. There are generally no withdrawal effects from St. John's wort, so you can stop and restart as needed. After a few months, it's a good idea to assess if you still need St. John's wort and at what dosage.

To do this, taper your dosage gradually, as opposed to stopping all at once.

Q. Can I combine St. John's wort with other herbs or supplements?

A. Herbs that are adaptogens can boost the positive effects of St. John's wort. (For a discussion of adaptogens, see Chapter 5.) Also, kava, with its quicker action, is useful for handling anxiety until the antidepressant effect of St. John's wort kicks in.

Q. Can I mix St. John's wort with 5-hydroxytryptophan (5-HTP)?

A. Both St. John's wort and 5-HTP increase the level of serotonin and can complement each other. However, watch for signs of too much serotonin. While rare, this serotonin syndrome has the following symptoms—a dangerous rise in blood pressure, diarrhea, fever, severe anxiety, headache, muscle tension, and confusion. The first sign is often a severe, throbbing headache. Stop both immediately and seek medical attention.

Q. Are there any risks in combining St. John's wort and other antidepressants?

A. So far, no adverse effects have been reported when St. John's wort has been combined with other antidepressants. However, when combining agents that act on serotonin, there is a risk of serotonin syndrome. Thus, medical supervision is essential if you are combining St. John's wort this way. Be alert for the signs of serotonin syndrome and immediately report them to your physician.

Q. I am currently taking Prozac and want to switch to St. John's wort. How can I do this?

A. First, reread the previous answer. Switching from Prozac to St. John's wort must be done under your doctor's supervision, with a gradual reduction of your medication and a gradual introduction of the St. John's wort over a period of several weeks. When changing from an MAO-inhibiting antidepressant drug to St. John's wort, there could be a dangerous rise in blood pressure when the two are mixed. If this happens, you may require a four-week "washout" period between stopping the drug and starting the herb.

9.

Saw Palmetto

Saw palmetto (*Serenoa repens*) is actually an extract of the saw palmetto berry, the fruit of a short palm tree that grows in the southeastern United States, mainly in Florida and Georgia. It is used to treat benign prostatic hyperplasia, or simply prostate enlargement. A traditional Native American remedy for urinary tract problems, saw palmetto was researched in the 1960s by French scientists, who developed the extract.

Q. How has saw palmetto been used?

A. Native Americans used saw palmetto berries in the treatment of various urinary problems in men, as well as for breast disorders in women. European and American physicians used saw palmetto as a treatment for benign prostatic hypertrophy (BPH), but in the United States, the herb's usage declined, along with that of all herbs, until a modern resurgence of interest and research. Saw palmetto is the main treatment for benign prostatic hyperplasia throughout Europe and, lately, the United States as well. There is excellent research to back up this use. Saw palmetto is also widely used to treat chronic prostatitis, or inflammation of the prostate.

Q. What is benign prostatic hypertrophy?

A. Benign prostatic hypertrophy, or BPH, is a benign (noncancerous) enlargement of the prostate gland. The condition affects at least 10 percent of men by the age of forty, and 50 percent of men by the age of fifty. The older a man is, the more likely he is to develop BPH.

The prostate gland is involved in the production of seminal fluid. More important here, though, is that this gland, which is the size of a walnut, surrounds the urethra, the tube through which the urine flows from the bladder. As the prostate enlarges, it narrows the urethral passageway, impairing the flow of the urine and causing a host of other problems. The typical symptoms of BPH include trouble starting urination, straining, a weak urinary stream, frequent urination, dribbling after urination, a sensation of incomplete emptying, and waking up several times at night to urinate. More serious problems include repeated bladder infections, involuntary urination, and bleeding.

Q. What is the usual treatment for BPH?

A. The prescription drugs Proscar (finasteride) and Hytrin (terazosin) are big business. A year's supply of Proscar costs $800. Saw palmetto costs a fraction of that. The old treatment was surgery, specifically transurethral resection of the prostate (TURP). In this surgery, the excess prostate tissue surrounding the urethra was cut away, enlarging the urethral opening, sort of like Roto-Rooter for the urethra. While the drugs are an improvement over this painful procedure, the herb is better yet.

Q. What results can I expect from saw palmetto?

A. About 90 percent of men respond to saw palmetto to some extent, beginning after approximately four to six weeks of treatment. Furthermore, while the prostate tends to continue growing when left untreated, saw palmetto causes a small but definite shrinkage. In other words, the herb does not simply relieve the symptoms, but may actually stop the prostate enlargement.

There have been many double-blind studies comparing the benefits of saw palmetto with a placebo, and the results have been excellent. The herb significantly improved the urinary-flow rate and other symptoms of prostate disease. Moreover, the response is rapid. For example, in one month-long study involving 110 patients taking 320 mg of saw palmetto daily, there was a significant increase in urinary flow and a decrease in night-time urination.

Q. How does saw palmetto work in treating BPH?

A. The male hormone testosterone is changed in the prostate to a more active form called dihydrotestosterone (DHT), which is a major cause of BPH. Saw palmetto works in two ways. It blocks 5-alpha-reductase, the enzyme that causes this change, and further blocks the binding of DHT to any cells.

Q. How well does saw palmetto combine with other herbs?

A. Saw palmetto is often combined with other herbs, including nettles (*Urtica dioica*) and pygeum (*Pygeum africanum*). A German study of 2,080 patients combined saw palmetto with nettles over a twelve-week period, with a significant improvement in symptoms. The subjects were given 160 mg of saw palmetto and 120 mg of nettles twice daily.

In another study, 250 men used either 50 mg of pygeum twice daily or a placebo along with the saw palmetto. The pygeum group did twice as well, with 66 percent showing improvement as compared to 33 percent of the placebo group.

Q. Is saw palmetto just as good as Proscar?

A. A recent double-blind study followed 1,098 men who took either saw palmetto or Proscar for six months. The treatments were equally effective, but while the Proscar caused impotence in some of the men, the saw palmetto caused no significant side effects. Proscar also lowers the PSA level, so its use may have the unintended effect of masking prostate cancer. Saw palmetto, on the other hand, leaves the PSA level unchanged, making it safer in this regard. Other studies have shown that saw palmetto is about as effective as the drugs alfuzosin and terazosin (Hytrin), and without their side effects.

Q. What is the recommended dose of saw palmetto?

A. The standard dose of saw palmetto is 160 mg twice a day of an extract standardized to contain 85- to 95-percent fatty acids and sterols, which are the active ingredients in the herb. A once-a-day dose of 320 mg may work just as well. The recommended dose of saw palmetto tincture is 2 to 5 ml three times daily of a concentration of one part herb to five parts alcohol.

Q. Does saw palmetto have any side effects?

A. The side effects of saw palmetto, which are rare, usually consist of mild digestive disturbance. Proscar, on the other hand, can cause impotence, decreased libido, and impaired sexual functioning. Saw palmetto improves sexual function. Which would you prefer?

Q. Can saw palmetto also be used by women?

A. Yes. There are a number of conditions in women, including some types of acne, excessive hair growth (hirsuitism), and fibrocystic breasts, that are the result of increased male-hormone production. Saw palmetto has the same action in women as it does in men of lowering the DHT level.

10.

How to Buy and Use Herbal Medicines

Selecting an herbal product—a tablet, capsule, tincture, or raw herb—can be a confusing experience. Different brands stress different features and benefits, and consumers often feel helpless while trying to sort out all the information. In this chapter, you will learn about the different forms in which herbal medicines are sold and how to shop for quality products.

Q. What forms do herbs come in, and which are best?

A. Herbs can be purchased as teas, tinctures, tablets, and capsules. Teas and tinctures, being liquid, may be absorbed by the body more rapidly than the other forms. Many herbalists recommend the liquid forms because, in tasting the herb, we begin the process of allowing it to heal us. Tablets and capsules are made from measured amounts of herb, and are the most common and convenient forms. Gelatin or vegetable-based capsules filled with powdered dried herb come in a variety of sizes and strengths, so you need to read the labels to ensure the proper dose. Tablets are powdered herb compressed into a solid pill, often with a variety of inert ingredients as fillers.

Q. How is tea prepared?

A. Tea is prepared from the whole herb, found in dried form in dark-colored glass jars at specialty herb shops. Shades of the old-fashioned apothecary! You take these dried herbs home and make them into teas, decoctions, or infusions.

Chinese medicine often uses decoctions. To make a decoction, boil a combination of dried herbs for a while to extract the medicine and reduce the liquid, thereby concentrating the tea. A weaker tea is called an infusion, made the way we make tea to drink, from tea leaves or tea bags. Pour boiling water over the herb, let it steep, strain the liquid (or remove the tea bag), and then drink the mixture. The common soothing herb chamomile is often prepared this way.

Q. What is a tincture?

A. A tincture is made by soaking the herb in alcohol. Some tinctures are made with glycerin to avoid the alcohol taste, but the resulting extract is weaker. If you prefer not to ingest alcohol, put the tincture in warm water or tea for a few minutes and let the alcohol evaporate. Doing this also disguises any remnant of the alcohol taste.

Q. What should I look for on the label to make sure it's the correct plant in the right amount?

A. The first thing to look for is the common name of the herb—for example, St. John's wort—followed by the botanical name—in this case, *Hypericum*. Next, look for the amount of herb in each unit—that is, capsule, tablet, or dropper—in grams (g or gm) or milligrams (mg). If the product is an extract, which is a concentrated form, the label should show the percent of the constituent to which the herb is standardized—for example, hypericin or hyperforin for St. John's wort. Ginkgo is standardized to ginkgolides, and kava to kavalactones.

Q. What is a standardized extract?

A. Unlike synthetic drugs, which contain a single

compound, herbs often have a variety of active ingredients. Therefore, we need to have a way of standardizing the products made from herbs—that is, to have a consistent, measured amount of product per unit dose, be it a capsule, tablet, or tincture. One ingredient, usually the one considered to be the active ingredient, is selected as the reference point, or marker. Even when the compound turns out not to be the active ingredient, it is often kept as the marker for convenience.

For example, hypericin was initially considered to be the main active ingredient of St. John's wort. However, later research indicated that it does not provide the major antidepressant activity. The hypericin content is still used as the marker, however, when creating standardized extracts. While other ingredients in St. John's wort may also be involved in the herb's antidepressant action, they are probably distributed within the plant in a way similar to that of hypericin. As a result, the hypericin standardization serves as a useful guidepost for the strength of all of the active ingredients in a St. John's wort product.

Q. How do manufacturers ensure that every batch of a product is the same?

A. Plants grow naturally rather than being manufactured. Thus, they vary in their content of active and marker ingredients. This is due to many variable factors—where the plant was grown, when it was grown, the season in which it was harvested, and even the time of day it was harvested. Good companies adjust their mixtures to account for these variations and to ensure standardized products.

Q. How do I know the correct dose for me?

A. The label usually gives the average dose. A ginkgo label, for example, might say one 60-mg tablet twice daily. This is meant to be a guideline, based on research and clinical use. While such a dose might improve memory, twice that dose is needed to treat Alzheimer's disease.

I recommend that you start with a relatively low dose, watch for a response, including unwanted effects, and adjust the dose accordingly. I have had patients who did well on 300 mg of St. John's wort once a day, while others needed four times that dose. Most fall in the middle, with the recommended 300 mg three times daily.

Q. Why doesn't the label tell you for what the herb should be used and what side effects you may encounter?

A. Most herbal products are regulated as dietary supplements. In 1994, the FDA's Dietary Supplement Health and Education Act (DSHEA) set new guidelines regarding the quality, labeling, packaging, and marketing of supplements. DSHEA allows manufacturers to make "statements of nutritional support for conventional vitamins and minerals," but since herbs aren't nutritional in the conventional sense, DSHEA allows them to make only what they call "structure and function claims."

Q. What can be said on the label?

A. The label can explain how the vitamin or herb affects the structure or function of the body. However, it can make no therapeutic or prevention claims, such as, "Treats headaches fast," or, "Cures the common cold." A saw palmetto label can say, "Helps maintain urinary and prostate health in men fifty and over." But it cannot say, "Helps treat the symptoms associated with benign prostatic hypertrophy," which is the actual reason men use it. That would mention the condition and the treatment, and be considered a drug claim.

Conclusion

Herbal medicines are effective and safe natural remedies. They generally have fewer side effects than drugs and are also relatively inexpensive. Scientific studies have provided a wealth of information about how they work. They are anything but unproven folklore. They work—I benefit from them, and so do my patients and countless other people.

If you've read this book from cover to cover, you now have a basic understanding of how some of the most popular medicinal herbs work and how to select them for your own use. They are nature's gift, working with our own bodies' chemistry and energy to promote healing and optimal health. They can hold their own against pharmaceuticals, and even do many things that pharmaceuticals cannot do.

Since our natural resources are dwindling, it's important to remember that with the widening use of herbs, we must be sure to replant and renew. There is no sense in using these miraculous products to promote our own health while interfering with the health of the planet. Moreover, any damage we do to nature comes right back to haunt us. As we destroy the rainforests, for example, we compromise our oxygen supply, literally choking ourselves in the process. We also are losing, irrevocably, hundreds of medicinal plants daily in this destruction.

My final words are a reminder to honor your mother, the Earth, and to walk lightly on her surface.

Glossary

Adaptogen. An herb that helps the body adapt to stresses of various kinds, causes no side effects, is effective at treating a wide variety of illnesses, and helps return the body to balance.

Alzheimer's disease. A condition that generally increases with age and causes increasing impairment of mental function. Also called senile dementia.

Antibody. A specific protein molecule made by a blood cell in response to a specific invader or antigen, such as a bacteria or virus.

Antioxidant. A substance that contributes an electron to a damaging free radical, making the free radical harmless.

Benign prostatic hypertrophy (BPH). A benign enlargement of the prostate gland that affects 50 percent of men by the age of fifty.

Free radical. An unpaired electron produced by the body as a result of metabolism and the toxic materials we ingest from our food, air, and water. Free radicals cause damage to the cells, and their effect increases with age. They probably are a contributing cause of cancer, heart disease, and aging.

Herb. As used in this book, a botanical medicine, botanical, or medicinal plant from any source.

Immune system. The complex combination of responses that fights invaders such as bacteria and viruses.

Intermittent claudication. A painful condition caused by poor circulation in the legs due to hardening of the arteries. It can interfere with walking.

Marker. A compound used as a convenient reference point when creating standardized extracts. The marker ensures that you get a specific amount of activity every time you buy a certain dosage unit.

Neurotransmitter. A chemical messenger of the brain. Serotonin, norepinephrine, and dopamine are the best known neurotransmitters, and deficiencies in these chemicals produce depression.

References

Bach D, et al., "Phytopharmaceutical and synthetic agents in the treatment of benign prostatic hyperplasia (BPH)," *Phytomedicine* 3 (1997):309–313.

Blumenthal M, Gruenwald J, et al., *The German Commission E Monographs*, English translation (Austin, TX: American Botanical Council, 1998).

Braeckman J, "The extract of *Serenoa repens* in the treatment of benign prostatic hyperplasia: a multi-center open study," *Current Therapeutic Research* 55 (July 1994):776–785.

Hobbs C, *The Echinacea Handbook* (Portland, OR: Eclectic Medical Publications, 1989).

Kupin, et al., "New data on *Eleutherococcus*," *Proceedings of the Second International Symposium on Eleutherococcus, Moscow* (1984):294–300.

Linde K, Ramirez G, Mulrow CD, Pauls A, Weidenhammer W, Melchart D, "St John's wort for depression: an overview and meta-analysis of randomized clinical trials," *British Medical Journal* 313 (1996):253–258.

"Saw palmetto," *British Journal of Urology* 81 (1998):383–387.

Steinmetz KA, Kushi LH, et al., "Vegetables, fruit and colon cancer in the Iowa women's health study," *American Journal of Epidemiology* 130 (1994):1–5.

Volz HP, et al., "Kava-kava extract WS 1490 versus placebo in anxiety disorders: a randomized placebo-controlled 25 week outpatient trial," *Pharmacopsychiatry* 30 (1997):1–5.

Warnecke G, "Psychosomatic dysfunctions in the female climacteric: clinical effectiveness and tolerance of kava extract WS 1490," *Fortschritte der Medizin* 109 (1991):119–122.

Suggested Readings

Balch JF and Balch PA. *Prescription for Nutritional Healing*, second edition. Garden City Park, NY: Avery Publishing Group, 1997.

Cass H. *Kava: Nature's Answer to Stress, Anxiety, and Insomnia*. Rocklin, CA: Prima Publishing, 1998.

Cass H. *St. John's Wort: Nature's Blues Buster*. Garden City Park, NY: Avery Publishing Group, 1998.

Foster S. *Herbs for Your Health*. Loveland, CO: Interweave Press, 1996.

Murray M and Pizzorno J. *Encyclopedia of Natural Medicine*. Rocklin, CA: Prima Publishing, 1994.

Schultz, Hansel, and Tyler. *Rational Phytotherapy: A Physician's Guide to Herbal Medicine*. New York: Springer, 1998.

Index